MUSEUMS,
MONUMENTS
MEMORIALS
2025-2026

State-by-State Guide to Military History Museums in the USA

Compiled by
Sherry McKee

EGP

ISBN: 978-1-958407-36-3

Cover photo: *The National World War II Museum, New Orleans*
© *Mick J. Prodger*

Book design by designpanache

ELM GROVE PUBLISHING
San Antonio, Texas, USA
www.elmgrovepublishing.com
Elm Grove Publishing is a legally registered trade name of Panache Communication Arts, Inc.

MUSEUMS, & MONUMENTS MEMORIALS
2025-2026

DEDICATION

There are a number of people in my life that deserve a dedication. My husband is number ONE. Staff Sergeant Gareld L. McKee was a Cuban Missile Crisis as well as a Vietnam veteran. He loved his Marine Corps, his country and his family. Gary was an avid military collector. His father, Bill, was with the 42nd Rainbow Division in WWII. He grew up in a small town that believed in honor and loyalty to his nation. My son and I were blessed to have had that honor and loyalty too!

AUTHOR'S PREFACE

I was born in a very small town in southern Illinois. My family was a tight knit, loving one that had been in the area long before the civil war. I was active in church and school. Being the oldest, my father taught me the things a lot of country girls learn. My father and I took my first car engine apart at the age of 10 and I learned to weld on a huge Lincoln arc welder at 12.

Since my father drew up the plans for our new home, I learned to face stone, mix concrete, read house plans and what had to be done to build a house.

I attended Eastern Illinois University majoring in Home Economics Education. At the beginning of my Sophomore year, I met my husband in an English class. He was a Vietnam Marine just returning home. My eyes were opened to the horrible way he and others were being treated. I graduated with a Bachelor in Education, a Masters in Guidance and Counseling plus hours in a Specialist in Education Administration degree.

During my life, I have been in 48 states and too many countries to list. My jobs are just as varied. They started with sales person in a ladies clothing store at sixteen. The list continues with secretary, school bus driver, high school counselor, Walmart employee, mother and many more. One job with which I am proud is being one of the first two women in a Texaco refinery as a laborer.

I was fortunate to have had my husband, Gary, for over 43 years. I am blessed to now live with our son, Patrick, and his family in Mt. Prospect, Illinois. Patrick wanted me to get out of the house and do something. I joined Frisbie Senior Center in Des Plaines and began to travel. I needed more. Not being gifted in computer knowledge, I wanted to do something different. Because I have always loved to travel and my husband was a veteran, I decided to gather information on museums in my area and put them in a binder that anyone could use. Then I decided I should include all branches of the service and it expanded.

Thus, this book...

My plan is to update this guide every year. Please feel free to send any memorials, monuments, or museums that you have found in your travels, and which are not featured in this edition. I will try to keep the book current. You may contact me through my publisher with suggestions for additions, correction or comments.

I hope you all have fun and safe travels!

SM

PUBLISHER'S NOTE

This is not intended to be a complete list of every military museum in the United States. It is a well-curated guide to select museums, monuments and memorials which may be of particular interest to those with a passion for details and a thirst for historical knowledge—and especially to veterans.

From large and prestigious institutions, to smaller, less well-known collections, these organizations are well worth a visit. But—please note—many smaller museums are run by volunteers and access can be limited, while others are located on military bases or government property and restrictions sometimes apply. Always check the website and, if possible, call ahead to check on opening hours and local conditions.

All information has been checked and is correct, to the best of the author's knowledge, at the time of going to print, but information can change, and neither the author nor the publisher are responsible for any errors which may occur. If you find a listed museum is no longer open or accessible, please let us know by contacting us via our website. Similarly, if there's a great museum in your town that was left out, or you discover a hidden gem not included in this guide while on your travels, please let us know so it can be included in a future edition.

elmgrovepublishing.com

ALABAMA

Battleship *Alabama* (BB-60) Memorial Park
2703 Battleship Pkwy
Mobile, AL 36602
www.ussalabama.com
251-433-2703 or 800-426-4929
Includes:
USS *Alabama* (BB-60)
USS *Drum* (SS-228)

Tuskegee Airmen National Historical Museum
Moton Field
1616 Chappie James Ave
Tuskegee, AL 36083-2985
www.nps.gov
334-421-0340

United States Army Aviation Museum
6000 Novosel St
Fort Novosel, AL 36362
(Formerly Fort Rucker)
armyaviationmuseum.org
334-598-2508 or 334-255-1078
(call for restrictions)

ALASKA

Alaska Aviation Museum
4721 Aircraft Dr
Anchorage, AK 99502
alaskaairmuseum.org
907-248-5325

Alaska Veterans Museum
411 W 4th Ave, Suite 2A
Anchorage, AK 99501
www.alaskaveterans.org
907-677-8802

**Aleutian World War II National Historic Area
Visitor Center**
2716 Airport Beach Rd
Unalaska, AK 99692
nps.gov/aleu/planyourvisit/index.htm
907-581-9944

Kodiak Military History Museum
1100 Abercrombie Dr
Miller Point, Fort Abercrombie
Kodiak, AK 99615
kadiak.org/museum/museum.html
907-486-7015

Prince William Sound Museum
100 Kenai St, #688
Whittier, AK 99693
pwsmuseum.org
907-486-7015

ARIZONA

Pima Air and Space Museum
6000 E Valencia Rd
Tucson, AZ 85756
pimaair.org
520-574-0462

The Navajo Code Talkers Museum and Veterans' Center
Window Rock Navajo Tribal Park
Window Rock, AZ 86515
discovernavajo.com
928-357-6291

Titan Missile Museum
1580 W Duval Mine Rd
Green Valley, AZ 85614
titanmissilemuseum.org
520-934-1863

Tucson Military Vehicle Museum
6000 E Valencia Rd #26
Tucson, AZ 85756
www.tucsonmilitaryvehicle.org
520-653-0080

ARKANSAS

Arkansas Inland Maritime Museum
North Little Rock Riverfront Park
120 Riverfront Park Dr
North Little Rock, AR 72114
nlr.ar.gov
501-371-8320

Arkansas Inland Maritime Museum (cont'd.)
Includes:
>USS *Hoga* (YT-146)
>USS *Razorback* (SS-394)
>USS *Scorpion* (SSN-589) Memorial
>USS *Snook* (SS-279) Memorial

CALIFORNIA

Battleship USS *Iowa* (BB-61) Museum
>250 S Harbor Blvd (Berth 87)
>(Los Angeles) San Pedro, CA 90731
>**pacificbattleship.com**
>310-971-4462 or 877-446-9261

Maritime Museum of San Diego
>1492 N Harbor Dr
>San Diego, CA 92101
>**www.sdmaritime.org**
>619-234-9153
Includes: USS *Dolphin* (AGSS-555)

Palm Springs Air Museum
>745 N Gene Autry Trail
>Palm Springs, CA 92262
>**palmspringsairmuseum.org**
>760-778-6262

Planes of Fame Museum
>14998 Cal Aero Dr
>Chino, CA 91710-9085
>**planesoffame.org**
>909-597-3722

SS *Jeremiah O'Brien*
Fisherman's Wharf (Pier 45)
San Francisco, CA 94133
www.ssjeremiahobrien.org
415-544-0100

SS *Lane Victory*
2400 Miner St (Berth 52)
(Los Angeles) San Pedro, CA 90731
www.lanevictory.org
310-519-9188

United States Navy Seabee Museum
3201 S Ventura Rd
Port Hueneme, CA 93043
seabee.org>heritage
805-982-5165

USS *Hornet* (CV-8) Museum
707 W Hornet Ave
Alameda Point (Pier 3)
Alameda, CA 94501
www.uss-hornet.org
510-521-8448

USS LCI (L)-1091
Humboldt Bay Maritime Museum
Association
Humboldt Bay Harbor (Adjacent to Berth 1)
Redwood Terminal
Samoa, CA 95564
humboldtbaymaritimemuseum.com
707-499-7366

USS LCS (L)(3)-102
Mare Island
Waterfront Ave and A Street
Vallejo, CA 94592
www.mightymidgets.org
415-661-9279

USS *Lucid* (MSO-458)
Building Futures Academy
3100 Monte Diablo Ave
Stockton, CA 95203
stocktonhistoricalmaritimemuseum.org
877-285-8243

USS *Midway* (CV-41) Museum
910 N Harbor Dr
San Diego, CA 92101
www.midway.org
619-544-9600

USS *Pampanito* (SS-383)
Fisherman's Wharf (Pier 45)
San Francisco, CA 94133
www.maritime.org
415-775-1943

COLORADO

4th Infantry Division Museum
6013 Nelson Blvd
Fort Carson, CO 80913
https://history.army.mil
719-524-0915

National Museum of World War II Aviation
755 Aviation Way
Colorado Springs, CO 80916-2740

www.worldwariiaviation.org
719-637-7559

Peterson Air & Space Museum
150 E Ent Ave
Peterson SFB, CO 80914-1303
petemuseum.org
719-556-5543

Pueblo Weisbrod Aircraft Museum
31001 Magnuson Ave
Pueblo, CO 81001
pwam.org
719-948-9219

Wings Over the Rockies Air and Space Museum
7711 E Academy Blvd
Denver, CO 80230-6929
wingsmuseum.org
303-360-5360

CONNECTICUT

National Coast Guard Museum *(Opening 2026)*
78 Howard St Suite A
New London, CT 06320
cgmuseumassociation.org
860-443-4200

New England Civil War Museum
14 Park Place
Vernon, CT 06066
newenglandcivilwarmuseum.com
860-870-3563

Submarine Force Library and Museum
1 Crystal Lake Rd
Groton, CT 06340
ussnautilus.org
860-343-0079
Includes: USS *Nautillus*

The Custom House Maritime Museum
150 Bank St
New London, CT 06320-6002
nlmaritimesociety.org
860-447-2501

West Haven Veterans Museum and Learning Center
30 Hood Terrace
West Haven, CT 06516
whmilmuseum.org
203-934-1111

DELAWARE

Air Mobility Command Museum
1301 Heritage Rd
Dover AFB, DE 19902
amcmuseum.org
302-677-5938 *(call for restrictions)*

Delaware Military Museum
1st Regiment Rd
Wilmington, DE 19808
delawaremilitarymuseum.org
302-751-5906

Fort Miles Museum
15099 Cape Henlopen Dr
Lewes, DE 19958
fortmilesmuseum.org
302-644-5007

FLORIDA

Brevard Veterans Memorial Center and Park
Library Plaza
400 S Sykes Creek Pkwy
Merritt Island, FL 32952
veteranmemorialcenter.org
321-453-1776

Museum of Military History
5210 W Irlo Bronson Memorial Hwy
Kissimmee, FL 34746
museumofmilitaryhistory.com
407-507-3894

National Naval Aviation Museum
1750 Radford Blvd
Pensacola, FL 32508
navalaviationmuseum.org
850-452-8450

National Navy UDT – Seal Museum
3300 N Hwy A1A
Hutchinson Island
Fort Pierce, FL 34949
www.navysealmuseum.org
772-595-5845

Road To Victory Military Museum
319 SE Stypmann Blvd
Stuart, FL 34994-2238
www.discovermarton.com
772-210-4283

Veterans Memorial Island Sanctuary
3001 Riverside Park Dr
Vero Beach, FL 32963-1874
covb.org
772-567-2144

Vietnam War Exhibit and Education Center
2475 Jen Dr, Ste 5
Melbourne, FL 32940
www.vietnamwarexhibit.com
321-212-9726

GEORGIA

Commemorative Air Force Airbase Georgia Museum
1200 Echo Court
Peach Tree City, GA 30269
airbasegeorgia.org
678-364-1110

Currahee Military Museum
160 N Alexander St
Toccoa, GA 30577
www.toccoahistory.com
706-282-5055

Museum of Aviation
1942 Heritage Blvd
Robins AFB, GA 31098
museumofaviation.org
478-926-6870
Includes: Georgia Aviation Hall of Fame

National Infantry Museum
1775 Legacy Way
Columbus, GA 31903
nationalinfantrymuseum.org
706-685-5800

National Museum of the Mighty 8th Air Force
175 Bourne Ave
Pooler, GA 31322
www.mightyeighth.org
912-748-8888

National Prisoner of War Museum
496 Cemetery Rd
Andersonville, GA 31711
georgiawwiitrail.org
229-924-0343

St. Marys Submarine Museum
102 St Marys St W
St Marys, GA 31558
georgiawwiitrail.org
912-882-2782

Webb Military Museum
411 E York St
Savannah, GA 31401
webbmilitarymuseum.com
912-663-0398

WWII Flight Training Museum
3 Airport Circle
Douglas, GA 31535
georgiawwiitrail.org
912-383-9111

WWII Home Front Museum
4201 1st St
St. Simons Island, GA 31522
georgiawwiitrail.org
912-634-7098

HAWAII

Battleship *Missouri* (BB-63) Memorial
Ford Island, Pearl Harbor
63 Cowpens St
Honolulu, HI 96818
ussmissouri.org
808-455-1600

Pearl Harbor Aviation Museum
Ford Island, Pearl Harbor
319 Lexington Blvd
Honolulu, HI 96818
www.pearlharboraviationmuseum.org
808-441-1000

USS *Arizona* (BB-39) Memorial
Pearl Harbor
1 Arizona Pl
Honolulu, HI 96818
www.pearlharborhistoricsites.org
808-422-3399 *(reservations recommended)*

USS *Bowfin* Submarine (SS/AGSS-287) Museum and Park
11 Arizona Memorial Dr
Honolulu, HI 96818
www.bowfin.org
808-423-1341

USS *Oklahoma* (BB-37) Memorial
Ford Island, Pearl Harbor
1 Arizona Pl
Honolulu, HI 96818
pearlharbor.org
808-422-3399

USS *Utah* (BB-31/AG-16) Memorial
Ford Island, Pearl Harbor
1275 Saratoga Rd
Honolulu, HI 96818-5029
www.ussutah1941.org
808-800-2337

IDAHO

Idaho Military Museum
4692 W Harvard St
Boise, ID 83705
museum.mil.idaho.gov
208-272-4841

Warhawk Air Museum
201 Municipal Dr
Nampa, ID 83687
warhawkairmuseum.org
208-465-6446

ILLINOIS

Air Combat Museum
Abraham Lincoln Capital Airport
835 S Airport Dr
Springfield, IL 62707-8486
aircombatmuseum.com
217-522-2181

First Division Museum at Cantigny
1s151 Winfield Rd
Wheaton, IL 60189-3353
www.fdmuseum.org
630-668-5161

Griffin Museum of Science and Industry – U-505 (German U-Boat)
5700 S DuSable
Lake Shore Dr
Chicago, IL 60637
www.msichicago.org
773-684-1414
(Submarine location: 1785 East Museum Dr.)

Heritage in Flight Museum
Logan County Airport
1351 Airport Rd
Lincoln, IL 62656
www.heritage-in-flight.org
217-732-3333 or 217-953-4118
(by appointment only)

Illinois Aviation Museum
110 S Clow International Pkwy
Bolingbrook, IL 60490
www.illinoisaviationmuseum.org
630-771-1937

Illinois Holocaust Museum and Education Center
9603 Woods Dr
Skokie, IL 60077
www.ilholocaustmuseum.org
847-967-4800

Illinois State Military Museum (National Guard)
1301 N MacArthur Blvd
Springfield, IL 62702-2399
militaryaffairs.illinois.gov
217-761-3910

Livingston County War Museum
321 N Main St
Pontiac, IL 61764
livingstoncountywarmusuem.com
815-842-0301

National Museum of the American Sailor
2531 Sheridan Rd, Bldg 42
Great Lakes, IL 60088
www.history.navy.mil
847-688-3154

National Veterans Art Museum
4041 N Milwaukee Ave, 2nd Floor
Chicago, IL 60641
www.nvam.org
312-683-9778

Naval Air Station Glenview Museum
2040 Lehigh Ave
Glenview, IL 60026
www.thehangarone.org
847-657-0000

Prairie Aviation Museum
2929 E Empire St
Bloomington, IL 61704
prairieaviationmuseum.org
309-663-7632

Rock Island Arsenal Museum
3500 N Ave, Bldg 60
Rock Island, IL 61299
arsenalhistoricalsociety.org
309-782-0485 (call for restrictions)

Russell Military Museum
43363 N US Hwy 41
Zion, IL 60099
www.russellmilitarymuseum.com
847-395-7020

Simpkins Military History Museum
605 E Cole St
Heyworth, IL 61745
www.enjoyillinois.com
390-319-3413

Vermilion County War Museum
307 N Vermilion St
Danville, IL 61832
vcwm.org
217-431-0034

INDIANA

Evansville Wartime Museum
7503 Petersburg Rd
Evansville, IN 47725
www.evansvillewartimemuseum.org
812-424-7461

Grissom Air Museum
1000 W Hoosier Blvd
Peru, IN 46970
www.grissomairmuseum.com
765-689-8011

Indiana Military Museum
715 S 6th St
Vincennes, IN 47591
www.indianamilitarymuseum.com
812-882-1941

Indiana War Memorial Museum
Indiana War Memorial Plaza
Historic District
55 E Michigan St
Indianapolis, IN 46204
www.in.gov
317-232-7615

Includes:
Indiana Gold Star Families Memorial
Indiana War Memorial
Indiana WWII Memorial
Korean War Memorial
Medal of Honor Memorial
9/11 Memorial
Soldiers and Sailors Monument
USS *Indianapolis* (CA-35) Memorial
Vietnam War Memorial

Museum of the Soldier, Inc.
510 E Arch St
Portland, IN 47371
museumofthesoldier.info
260-726-2967

IOWA

Iowa Aviation Museum
2251 Airport Rd
Greenfield, IA 50849
flyingmuseum.com
641-343-7184

Iowa Gold Star Military Museum
7105 NW 70th Ave
Johnston, IA 50131
goldstarmuseum.iowa.gov
515-252-4531

KANSAS

Combat Air Museum
7016 SE Forbes Ave
Forbes Field, Topeka, KS 66619-1444
combatairmuseum.org
785-862-3303

Museum of the Kansas National Guard
125 SE Airport Dr
Topeka, KS 66619
kansasguardmuseum.com
785-862-1020

U.S. Cavalry Museum
205 Henry Ave
Fort Riley, KS 66442
https://home.army.mil/riley/about/museums
785-239-2737

KENTUCKY

Frazier History Museum
829 W Main St
Louisville, KY 40202
fraziermuseum.org
502-753-5663

General George Patton Museum of Leadership
4554 Fayette Ave
Fort Knox, KY 40121-0208
generalpatton.org
502-624-3391

Kentucky Military History Museum
125 E Main St
Frankfort, KY 40601
history.ky.gov
502-564-1792

Kentucky Veteran and Patriot Museum
635 Phillips Dr
Wickliffe, KY 42087
kvpm.org
270-210-2452 *(call for hours)*

LOUISIANA

Chennault Aviation & Military Museum
701 Kansas Ln
Monroe, LA 71203
chennaultmuseum.org
318-362-5540

Louisiana Maneuvers & Military Museum
623 G St
Pineville, LA 71360
geauxguardmuseums.com
318-290-5733

Regional Military Museum
1154 Barrow St
Houma, LA 70360
regionalmilitarymuseum.com
985-873-8200

The National World War II Museum
945 Magazine St
New Orleans, LA 70130-0208
www.nationalww2museum.org
504-528-1944

USS *Kidd*
305 South River Rd
Baton Rouge, LA 70802
usskidd.com
225-342-1942

MAINE

Maine Armed Forces Museum
194 Winthrop St
Augusta, ME 04330
mainearmedforcesmuseum.org
207-649-5989

Maine Military Museum
50 Peary Terrace
South Portland, ME 04106
mainemilitarymuseum.org
207-767-8227

MARYLAND

Maryland Veterans Museum at Patriot Park
11000 Crain Hwy N
Newburg, MD 20664
www.mdvets.cc
301-932-1900

**The Maryland Museum of Military History
and Historical Research Center**
Fifth Regiment Armory
29th Division St
Baltimore, MD 21201
military.maryland.gov
667-296-3470 *(call for hours)*

United States Naval Academy Museum
118 Maryland Ave
Annapolis, MD 21402
www.usna.edu/Museum
410-293-2108

MASSACHUSETTS

American Heritage Museum
568 Main St
Hudson, MA 01749
www.americanheritagemuseum.org
978-562-9182

**America's Fleet Museum —
Battleship Cove**
5 Water St
Fall River, MA 02721
www.battleshipcove.org
508-678-1100

Battleship Cove (cont'd.)
Includes:
>Gimik "Gizmo" (OSS semi-submersible)
>LCM 56 (Landing Craft Mechanized)
>PT 617
>PT 796
>USS *Joseph P. Kennedy Jr.* (DD-850)
>>*(closed for restoration)*
>USS *Lionfish* (SS-298)
>>*(closed for restoration)*
>USS *Massachusetts* (BB-59)

Custom House Maritime Museum
>25 Water St
>Newburyport, MA 01950-2754
>**customhousemaritimemuseum.org**
>978-462-8681

Fort Devens Museum
>94 Jackson Rd, #305
>Devens, MA 01434
>**fortdevensmuseum.org**
>978-772-1286

Military Museum and Armory
>1 Faneuil Hall Square, Fourth Floor
>Boston, MA 02109
>**faneuilhallmarketplace.com**
>617-227-1638

Springfield Armory
>Springfield Techincal Community College
>1 Armory Square
>Springfield, MA 01105
>**nps.gov/spar**
>413-734-8551

USS *Constitution* **Museum**
Bldg 22, Charleston Navy Yard
Charleston, MA 02129
ussconstitutionmuseum.org
617-426-1812

MICHIGAN

**Air Zoo Aerospace And Science Museum —
Flight Discovery Center**
3101 E Milham Rd
Kalamazoo, MI 49002
www.airzoo.org
269-382-6555 or 866-524-7966

**Air Zoo Aerospace and Science Museum —
Flight Innovation Center**
6151 Portage Rd
Portage, MI 49002
www.airzoo.org
269-382-6555 or 866-524-7966

Michigan Heroes Museum
1250 Weiss St
Frankenmuth, MI 48734
www.miheroes.org
989-652-8005

**Saginaw Valley Naval Ship Museum
USS** *Edson* **(DD-946)**
1680 Martin St
Bay City, MI 48706
svnsm.com
989-684-3946

The Tuskegee Airmen National Historical Museum
315 E Warren Ave
Detroit, MI 48201
tuskegeemuseum.org
313-843-8849

USS *Silversides* Submarine (SS-236) Museum
1346 Bluff St
Muskegon, MI 49441
silversidesmuseum.org
231-755-1230

World War II Glider and Military Museum
302 Kent St
Iron Mountain, MI 49801
www.menomineemuseum.com>glider
906-774-1086

Yankee Air Museum
Willow Run Airport
47884 D St
Belleville, MI 48111-1126
miflightmuseum.org
734-483-4030

MINNESOTA

Herreid Military Museum
213 E Luverne St
Luverne, MN 56156
herreidmilitarymuseum.org
507-283-4061

Minnesota Military and Veterans Museum
Camp Ripley
15000 Hwy 115

Little Falls, MN 56345-4173
mnvetmuseum.org
320-616-6050

MISSISSIPPI

**Beauvoir: The Jefferson Davis Home
and Presidential Library**
2244 Beach Blvd
Biloxi, MS 39531
visitbeauvoir.org
228-388-4400

G.I. Museum
5796 Ritcher Rd
Ocean Springs, MS 39564
gimuseum.com
228-872-1943

Mississippi Armed Forces Museum
Bldg 850, Forrest Ave,
Camp Shelby, MS 39407
msarmedforcesmuseum.org
601-558-2757

Tupelo Veteran's Museum
689 Rutherford Rd
Tupelo, MS 38801
tupeloveteransmuseum.com
662-844-1515

Mississippi Vietnam Veterans Memorial
3704 Bienville Blvd (US Hwy 90 E)
Ocean Springs, MS 39564
psysim.tripod.com/msvn
228-392-7190

MISSOURI

Branson Veterans Memorial Museum
1250 W Hwy 76 Country Blvd
Branson, MO 65616
veteransmemorialbranson.com
417-336-2300

National WWI Museum and Memorial
2 Memorial Dr
Kansas City, MO 64108
www.theworldwar.org
816-888-8100

United States Army Military Police Corps Regimental Museum
14296 S Dakota Ave
Fort Leonard Wood, MO 65473
mpraonline.org
573-329-6772 *(call for restrictions)*

MONTANA

Custer Battlefield Museum
I-90, exit 514
Garryowen, MT 59031
custermuseum.org
406-638-1876

Little Bighorn Battlefield National Monument
858 Little Bighorn Battlefield Rd
Garryowen, MT 59031
www.nps.gov
406-638-2621

Montana Military Museum
  1956 Mt. Majo St
  Fort Harrison, MT 59636
  montanamilitarymuseum.org
  406-324-3550

NEBRASKA

Civil War Veterans Museum
  910 1st Corso
  Nebraska City, NE 68410
  civilwarmuseumnc.org
  402-873-4018

Heartland Museum of Military Vehicles
  606 Heartland Rd
  Lexington, NE 68850
  heartlandmuseum.com
  308-324-6329

Nebraska National Guard Museum
  201 N 8th St
  Seward, NE 68434
  www.nengm.org
  402-309-8763

Sallows Military Museum
  1101 Niobrara Ave
  Alliance, NE 69301
  sallowsmilitarymuseum.com
  308-762-2385

Strategic Air Command & Aerospace Museum
  28210 W Park Hwy
  Ashland, NE 68003
  sacmuseum.org
  402-944-3100

NEVADA

Hawthorne Ordnance Museum
925 E St
Hawthorne, NV 89415
hawthorneordnancemuseum.com
775-945-5400

NEW HAMPSHIRE

Aviation Museum of New Hampshire
27 Navigator Rd
Londonderry, NH 03053
aviationmuseumofnh.org
603-669-4820

USS *Albacore* (AGSS-569)
Albacore Park
569 Submarine Way
Portsmouth, NH 03801
ussalbacore.org
603-436-3680

Wright Museum of WWII
77 Center St
Wolfeboro, NH 03894
wrightmuseum.org
603-569-1212

NEW JERSEY

Battleship *New Jersey* (BB-62) Museum and Memorial
> 62 Battleship Place
> Camden, NJ 08103
> **www.battleshipnewjersey.org**
> 856-966-1652
> 866-877-6262

Lawrenceville Armory
> 161 Eggerts Crossing Rd
> Lawrenceville, NJ 08648
> **njmilitiamuseum.org/lawrenceville**
> 609-671-2980

Military Technology Museum of New Jersey
> 2201 Marconi Rd
> Wall Township, NJ 07719
> **mtmnj.org**
> 848-404-9774

NAS Wildwood Aviation Museum
> 500 Forrestal Rd
> Cape May, NJ 08204
> **usnasw.org**
> 609-886-8787

National Guard Militia Museum of New Jersey
> 100 Camp Dr, Bldg 7
> Sea Girt, NJ 08750
> **njmilitiamuseum.org**
> 732-974-4570

Old Barracks Museum
101 Barrack St
Trenton, NJ 08608
barracks.org
609-396-1776

NEW MEXICO

Hamilton Military Museum
996 S Broadway St
Truth or Consequences, NM 87901
sierracountynewmexico.info>
575-894-0750

New Mexico Military Institute McBride Museum
101 W College Blvd
Roswell, NM 88201
www.nmmi.edu/mcbride-museum/
575-624-8384

New Mexico Military Museum
1050 Old Pecos Trail
Santa Fe, NM 87505
newmexicomilitarymuseum.com
505-476-1479

Navajo Code Talker Museum
106 W Historic Hwy 66
Gallup, NM 87301-6226
www.thegallupchamber.com
505-722-2228

War Eagles Air Museum
8012 Airport Rd
Santa Teresa, NM 88008
www.wareaglesairmuseum.com
575-589-2000

NEW YORK

American Merchant Marine Museum
U.S. Merchant Marine Academy
300 Steamboat Rd
Kings Point, NY 11024
www.usmma.edu
516-726-6047

National Soaring Museum
51 Soaring Hill Dr
Elmira, NY 14903
www.soaringmuseum.org
607-734-3128

USS *Growler* (SSG-577)
Intrepid (CV-11) Museum
W 46th St and 12th Ave (Pier 86)
New York, NY 10036
www.history.navy.mil
212-245-0072 or 877-957-7447

USS *Intrepid* (CV-11) Museum
W 46th St and 12th Ave (Pier 86)
New York, NY 10036
www.intrepidmuseum.org
212-245-0072 or 877-957-7447

NORTH CAROLINA

The 82nd Airborne Division War Memorial Museum
5108 Ardennes St, Bldg c-6841
Fort Bragg, NC 28310
82ndairbornedivisionmuseum.com
910-432-3443 or 910-432-5307

USS *North Carolina* (BB-55)
    1 Battleship Rd
    Wilmington, NC 28401
    www.battleshipnc.com
    910-399-9100

NORTH DAKOTA

Chateau de Mores State Historic Site
    3426 Chateau Rd
    Medora, ND 58645
    www.history.nd.gov>historicsites>chateau
    701-623-4355

Dakota Territory Air Museum
    100 34th Ave NE
    Minot, ND 58702-0195
    dakotaterritoryairmuseum.com
    701-852-8500

Fargo Air Museum
    1609 19th Ave N
    Fargo, ND 58102
    fargoairmuseum.org
    701-293-8043

North Dakota Heritage Center & State Museum
    612 E Boulevard Ave
    Bismarck, ND 58505
    statemuseum.nd.gov
    701-328-2666

Ronald Reagan Minuteman Missile Site
    555 113$\frac{1}{2}$ Ave NE
    Cooperstown, ND 58425
    www.history.nd.gov/historicsites/
    minutemanmissile/index
    701-797-3691

OHIO

National Museum of the United States Air Force
   Wright-Patterson AFB
   1100 Spaatz St
   Dayton, OH 45433
   www.nationalmuseum.af.mil
   937-255-3286

National Veterans Memorial and Museum
   300 W Broad St
   Columbus, OH 43215
   nationalvmm.org
   888-987-6866

OKLAHOMA

Fort Sill National Historic Landmark and Museum
   435 Quanah Rd
   Fort Sill, OK 73503-5100
   sill-www.army.mil
   580-442-5123

The United States Army Field Artillery Museum
   238 Randolph Rd
   Lawton, OK 73503
   sill-www.army.mil/famuseum
   580-442-1819

OREGON

Oregon Coast Military Museum
2145 Kingwood St
Florence, OR 97439
oregoncoastmilitarymuseum.com
541-902-5160

Oregon Military Museum
15300 SE Minuteman Way
Clackamas, OR 97015
oregonmilitarymuseum.org
971-355-2275

PENNSYLVANIA

Air Heritage Aviation Museum
35 Piper St, #1043
Beaver Falls, PA 15010
airheritage.org
724-843-2820

Museum of the American Revolution
101 S 3rd St
Philadelphia, PA 19106
www.amrevmuseum.org
215-253-6731

U.S. Army Heritage and Education Center Foundation
950 Soldiers Dr
Carlisle, PA 17013-5021
ahec.armywarcollege.edu
717-245-3972

RHODE ISLAND

Seabee Museum and Memorial Park
21 Iafrate Way
North Kingstown, RI 02852
seabeesmuseum.com
401-294-7233

The International Museum of World War II
344 Main St
Wakefield, RI 02879
wwiifoundation.org
401-862-3030

Varnum Memorial Armory Museum
6 Main St
East Greenwich, RI 02818
varnumcontinentals.org
401-884-4110

SOUTH CAROLINA

Hunley Museum
1250 Supply St
Charleston, SC 29405
www.hunley.org
843-743-4865

Patriots Point Naval & Maritime Museum
40 Patriots Point Rd
Mt. Pleasant, SC 29464
www.patriotspoint.org
843-884-2727
Includes:
USS *Laffey* (DD-724)
USS *Yorktown* (CV-10)

SOUTH DAKOTA

South Dakota Air and Space Museum
2890 Davis Dr
Ellsworth AFB, SD 57706
nps.gov/places/sdaasm.htm
605-385-5188

South Dakota National Guard Museum
301 E Dakota Ave
Pierre, SD 57501
sd.gov/ngm
605-773-2475

TENNESSEE

National Medal of Honor Heritage Museum
2 W Aquarium Way, Suite 104
Chattanooga, TN 37402
mohhc.org
423-877-2525

Sam H. Werner Military Museum
1148 W Main St
Monteagle, TN 37356
wernermilitarymuseum.com
931-607-9200

Tennessee Museum of Aviation
135 Air Museum Way
Sevierville, TN 37862
tnairmuseum.com
865-908-0171

The Veterans' Museum at Dyersburg Army Air Base
    100 Veterans Dr N
    Halls, TN 38040
    www.dyaab.us
    731-836-7400

Wilson County Veterans Museum
    304 E Main St
    Lebanon, Tennessee 37087
    wilsoncountyveteransmuseum.com
    615-444-2460

TEXAS

Army Air Corps Library and Museum
    822 Clemente
    Irving, TX 75039
    www.armyaircorpsmuseum.org
    214-957-1393

Commemorative Air Force Headquarters
    5661 Mariner Dr
    Dallas, TX 75237
    www.commemorativeairforce.org
    214-330-1700
CAF has numerous locations throughout the U.S.A.

Frontier of Flight Museum
    Love Field
    6911 Lemmon Ave
    Dallas, TX 75209
    www.flightmuseum.com
    214-350-3600

National Medal of Honor Museum
1861 AT&T Way
Arlington, TX 76011
mohmuseum.org
817-274-1861

National Mounted Warrior Museum
Fort Cavazos (Formerly Fort Hood)
Bldg 6900 105 Trooper Loop,
Fort Cavazos, TX 76544
history.army.mil/museums/NMWM
254-286-5684

National Museum of the Pacific War
311 E Austin St
Fredericksburg, TX 78624
www.pacificwarmuseum.org
830-997-8600

National WASP WWII Museum
210 Avenger Field Rd
Sweetwater, TX 79556
www.waspmuseum.org
325-235-0099

Texas Air Museum
Stinson Field
1234 99th St
San Antonio, TX 78214
www.texasairmuseum.org
210-977-9885

The National Vietnam War Museum
12685 Mineral Wells Hwy
Weatherford, TX 76088
www.nationalvnmuseum.org
682-239-0683

The Vietnam War Flight Museum
12101 Blume Ave
Houston, TX 77034
www.vietnamwarflight.com
832-377-5596

USS *Lexington* (CV-16)
2914 N Shoreline Blvd
Corpus Christi, TX 78402
www.usslexington.com
361-888-4873
800-LADYLEX (800-523-9539)

USS *Texas* (BB-35)
Gulf Copper Dry Dock and Rig Repair
2920 Todd Shipyard Rd
Galveston, TX 77554
battleshiptexas.org
832-841-3500 (Foundation)
Currently in limbo

UTAH

Fort Douglas Military Museum
32 Potter St
Salt Lake City, UT 84113
fortdouglas.org
801-581-1251

Hill Aerospace Museum
7961 Wardleigh Rd
Hill AFB, UT 84056
aerospaceutah.org
801- 825-5817

Historic Wendover Airfield
352 Airport Way E
Wendover, UT 84083
wendoverairfield.org
435-665-7724

John M. Browning Firearms Museum
Union Station
2501 Wall Ave, Ste A
Ogden, UT 84401-1359
ogdencity.gov/1333/browning-firearms
801-629-8680

Western Sky Aviation Warbird Museum
4196 S Airport Pkwy
St. George, UT 84790
westernskywarbirds.org
435-669-0655

VERMONT

Fort Ethan Allen Museum
11 Marcy Dr
Essex, VT 05452
parkinsonbooks.com>fortethanallen
museum
802-482-3113

Harriet Farnsworth Powell Historical Museum
3 Browns River Rd
Essex, VT 05452
essexcommunityhistoricalsociety.org
802-878-1354

Vermont National Guard Library and Museum
789 Vermont National Guard Rd
Colchester, VT 05446
https://vt.public.ng.mil/Museum/Library/
802-338-3360

VIRGINIA

Hampton Roads Naval Museum
1 Waterside Dr, Suite 248
Norfolk, VA 23510
**www.history.navy.mil/content/history/
museums/hrnm.html**
757-322-2987

Military Aviation Museum
1341 Princess Anne Rd
Virginia Beach, VA 23457
www.militaryaviationmuseum.org
757-721-7767

National Air and Space Museum
Steven F. Udvar-Hazy Center
14390 Air and Space Museum Pkwy
Chantilly, VA 20151
www.airandspace.si.edu
703-572-4118

National Museum of the Marine Corps
1775 Semper Fidelis Hwy
Triangle, VA 22172
www.usmcmuseum.com
703-432-1775, 800-397-7585
or 877-653-1775

National Museum of the United States Army
1775 Liberty Dr
Fort Belvoir, VA 22060
www.thenmusa.org
800-506-2672

National Women's History Museum
205 S Whiting St, Ste 401
Alexandria, VA 22304
womenshistory.org
703-461-1920

The Mariner's Museum and Park
100 Museum Dr
Newport News, VA 23606
www.marinersmuseum.org
757-596-2222

United States Army Women's Museum
2100 A Ave, Bldg 5219
Fort Gregg-Adams, VA 23801
awm.army.mil
804-734-4327

Virginia War Museum
9285 Warwick Blvd
Newport News, VA 23607
newportnewshistory.org
757-247-8523

WASHINGTON

Boeing Flight Aviation Center
8415 Paine Field Blvd
Mukilteo, WA 98275
www.boeingfutureofflight.com
800-464-1476

Heritage Flight Museum
15053 Crosswind Dr
Burlington, WA 98233
heritageflight.org
360-424-5151

Puget Sound Naval Museum
251 1st St
Bremerton, WA 98337
pugetsoundnavymuseum.org
360-479-7447

The Museum of Flight
9404 E Marginal Way S
Seattle, WA 98108-4097
www.museumofflight.org
206-764-5700

U.S. Naval Undersea Museum
1 Garnett Way
Keyport, WA 98345
navalunderseamuseum.org
360-396-4148

WEST VIRGINIA

Mercer County War Museum
1500 W Main St
Princeton, WV 24740
visitmercercounty.com
304-487-3670

Mountaineer Military Museum
345 Center Ave
Weston, WV 26452
www.mountaineermilitarymuseum.com
304-472-3943

WISCONSIN

Central Wisconsin Veterans Memorial Cemetery
N 2665 County Rd QQ
King, WI 54946
dva.wi.gov
715-256-5000

Military Veterans Museum
4300 Poberezny Rd
Oshkosh, WI 54902-2194
mvmec.org
920-426-8615

Milwaukee County War Memorial Center
750 N Lincoln Memorial Dr
Milwaukee, WI 53202
warmemorialcenter.org
414-273-5533

Northern Wisconsin Veterans Memorial Cemetery
N4063 Veterans Way
Spooner, WI 54801
dva.wi.gov
715-635-5360

Pritzker Military Museum and Library
10475 12th St
Kenosha, WI 53144
pritzkermilitary.org
262-800-7402

Richard I. Bong Veterans Historical Center
305 E 2nd St
Superior, WI 54880
bongcenter.org
715-392-7151

Soldiers Walk Veterans Memorial Park
551 Park Dr
Arcadia, WI 54612
www.soldierswalkmemorialpark.com
608-323-2319

Wisconsin National Guard Museum
Volk Field
101 Independence Dr
Camp Douglas, WI 54618
museumsusa.org
608-427-1280

Wisconsin Veterans Museum
30 West Mifflin St
Madison, WI 53703
wisvetsmuseum.com
608-267-1799

WYOMING

National Museum of Military Vehicles
6419 US26
Dubois, WY 82513
nmmv.org
307-455-3802

Warren ICBM & Heritage Museum
    401 Champagne Dr Bldg 31
    Francis E. Warren AFB, WY 82005
    www.warrenmuseum.com
    307-773-2980

Wyoming National Guard Museum
    624 E. Pershing Blvd
    Cheyenne, WY 82009
    www.wyomilitary.wyo.gov
    307-432-0057

DISTRICT OF COLUMBIA

Clara Barton Missing Soldiers Office Museum
    437 7th Street NW
    Washington, DC 20004
    clarabartonmuseum.org
    202-824-0613

Korean War Veterans Memorial
    5 Henry Bacon Drive NW
    Washington, DC 20037
    www.nps.gov/kowa
    202-426-6841

National Air and Space Museum (Smithsonian)
    6th St and Independence Ave SW
    Washington, DC 20560
    airandspace.si.edu
    202-633-2214

National Guard Memorial Museum
    1 Massachusetts Ave NW
    Washington, DC 20001
    www.ngaus.org
    202-789-0031

National Museum of American Jewish Military History
1811 R St NW
Washington, DC 20009
nmajmh.org
202-265-6280

National Museum of the U.S. Navy
736 Sicard St SE
Washington, DC 20374
www.history.navy.mil
202-685-0589

The United States Navy Memorial
801 Pennsylvania Ave NW
Washington, DC 20004-2608
www.navymemorial.org
202-737-2300

Vietnam Veterans Memorial
5 Henry Bacon Dr NW
Washington, DC 20002
www.nps.gov
202-426-6841

INDEX

Custom House Maritime Museum, Newburyport, MA 30
Dakota Territory Air Museum, Minot, ND 40
Delaware Military Museum, Wilmington, DE 16
Evansville Wartime Museum, Evansville, IN 24
Fargo Air Museum, Fargo, ND 40
First Division Museum at Cantigny, Wheaton, IL 22
Fort Devens Museum, Devens, MA 30
Fort Douglas Military Museum, Salt Lake City, UT 47
Fort Ethan Allen Museum, Essex, VT 48
Fort Miles Museum, Lewes, DE 17
Fort Sill National Historic Landmark and Museum, Fort Sill, OK 41
Frazier Kentucky History Museum, Louisville, KY 27
Frontier of Flight Museum, Dallas, TX 45
General George Patton Museum of Leadership, Fort Knox, KY 27
Georgia Aviation Hall of Fame, Warner Robins AFB, GA 19
G.I. Museum, Ocean Springs, MS 33
Gimik "Gizmo" (OSS semi-submersible), Fall River, MA 30
Griffin Museum of Science and Industry – U505 (German U-Boat),
      Chicago, IL 22
Grissom Air Museum, Peru, IN 25
Hamilton Military Museum, Truth or Consequences, NM 38
Hampton Roads Naval Museum, Norfolk, VA 49
Harriet Farnsworth Powell Historical Museum, Essex, VT 48
Hawthorne Ordnance Museum, Hawthorne, NV 36
Heartland Museum of Military Vehicles, Lexington, NE 35
Heritage Flight Museum, Burlington, WA 51
Heritage in Flight Museum, Lincoln, IL 22
Herreid Military Museum, Luverne, MN 32
Hill Aerospace Museum, Hill AFB, UT 47
Historic Wendover Airfield, Wendover, UT 48
Hunley Museum, Charleston, SC 43
Idaho Military Museum, Boise, ID 21
Illinois Aviation Museum, Bolingbrook, IL 22
Illinois Holocaust Museum and Education Center, Skokie, IL 23
Illinois State Military Museum (National Guard), Springfield, IL 23
Indiana Gold Star Families Memorial, Indianapolis, IN 25
Indiana Military Museum, Vincennes, IN 25
Indiana War Memorial, Indianapolis, IN 25
Indiana War Memorial Museum, Indianapolis, IN 25
Indiana WWII Memorial, Indianapolis, IN 25
Iowa Aviation Museum, Greenfield, IA 26
Iowa Gold Star Military Museum, Johnson, IA 26
John M. Browning Firearms Museum, Ogden, UT 48

PT796 (PT Boat), Fall River, MA 30
Pueblo Weisbrod Aircraft Museum, Pueblo, CO 15
Puget Sound Naval Museum, Bremerton, WA 51
Regional Military Museum, Houma, LA 28
Richard I. Bong Veterans Historical Center, Superior, WI 53
Road To Victory Military Museum, Stuart, FL 18
Rock Island Arsenal Museum, Rock Island, IL 24
Ronald Reagan Minuteman Missile Site, Cooperstown, ND 40
Russell Military Museum, Zion, IL 24
Saginaw Valley Naval Ship Museum, Bay City, MI 31
Sallows Military Museum, Alliance, NE 35
Sam H. Werner Military Museum, Monteagle, TN 44
Seabee Museum and Memorial Park, North Kingstown, RI 43
Simpkins Military History Museum, Heyworth, IL 24
Soldiers and Sailors Monument, Indianapolis, IN 25
Soldiers Walk Veterans Memorial Park, Arcadia, WI 53
South Dakota Air and Space Museum, Ellsworth AFB, SD 44
South Dakota National Guard Museum, Pierre, SD 44
Springfield Armory, Springfield, MA 30
SS *Jeremiah O'Brien*, San Francisco, CA 13
SS *Lane Victory*, San Pedro, CA 13
St. Marys Submarine Museum, St Marys, GA 19
Strategic Air Command & Aerospace Museum, Ashland, NE 35
Submarine Force Library and Museum, Groton, CT 16
Tennessee Museum of Aviation, Sevierville, TN 44
Texas Air Museum, San Antonio, TX 46
The 82nd Airborne Division War Memorial Museum, Fort Bragg, NC
     39
The Custom House Maritime Museum, New London, CT 16
The International Museum of World War II, Wakefield, RI 43
The Mariner's Museum and Park, Newport News, VA 50
The Maryland Museum of Military History and Historical Research
     Center, Baltimore, MD 29
The Museum of Flight, Seattle, WA 51
The National Vietnam War Museum, Weatherford, TX 46
The National World War II Museum, New Orleans, LA 28
The Navajo Code Talkers Museum, Window Rock, AZ 11
The Tuskegee Airmen National Historical Museum, Detroit, MI 32
The United States Army Field Artillery Museum, Lawton, OK 41
The United States Navy Memorial, Washington, DC 55
The Veterans' Museum at Dyersburg Army Air Base, Halls, TN 45
The Vietnam War Flight Museum, Houston, TX 47
Titan Missile Museum, Green Valley, AZ 11

Tucson Military Vehicle Museum, Tucson, AZ 11
Tupelo Veteran's Museum, Tupelo, MS 33
Tuskegee Airmen National Historical Museum, Tuskegee, AL 9
U505 (German U-Boat), Chicago, IL 22
United States Army Aviation Museum, Fort Novosel, AL 9
United States Army Military Police Corps Regimental Museum, Fort
     Leonard Wood, MO 34
United States Army Women's Museum, Fort Gregg-Adams, VA 50
United States Naval Academy Museum, Annapolis, MD 29
United States Navy Seabee Museum, Port Hueneme, CA 13
U.S. Army Heritage and Education Center Foundation, Carlisle, PA 42
U.S. Cavalry Museum, Fort Riley, KS 26
U.S. Naval Undersea Museum, Keyport, WA 51
USS *Albacore* (AGSS-569), Portsmouth, NH 36
USS *Arizona* (BB-39) Memorial, Honolulu, HI 20
USS *Bowfin* Submarine (SS/AGSS-287), Honolulu, HI 21
USS *Constitution* Museum, Charleston, MA 31
USS *Dolphin* (AGSS-555), San Diego, CA 12
USS *Drum* (SS-228), Mobile, AL 9
USS *Edson* (DD-946), Bay City, MI 31
USS *Growler* (SSG-577), New York, NY 39
USS *Hoga* (YT-146), North Little Rock, AR 12
USS *Hornet* (CV-8) Museum, Alameda, CA 13
USS *Indianapolis* (CA-35) Memorial. Indiana War Memorial Museum,
     Indianapolis, IN 25
USS *Intrepid* (CV-11) Museum, New York, NY 39
USS *Iowa* (BB-61), San Pedro, CA 12
USS *Joseph P. Kennedy, Jr* (DD-850), Fall River, MA 30
USS *Kidd*, Baton Rouge, LA 28
USS *Laffey* (DD-724), Mt. Pleasant, SC 43
USS LCI (L)-1091, Humboldt Bay Maritime Museum Association
     Humboldt Bay Harbor, Samoa, CA 13
USS LCS (L)(3)-102, Vallejo, CA 14
USS *Lexington* (CV-16), Corpus Christi, TX 47
USS *Lionfish* (SS-298), Fall River, MA 30
USS *Lucid* (MSO-458)Building Futures Academy, Stockton, CA 14
USS *Massachusetts* (BB-59), Fall River, MA 30
USS *Midway* (CV-41) Museum, San Diego, CA 14
USS *Nautilus* (SSN-571), Groton, CT 16
USS *North Carolina* (BB-55), Wilmington, NC 40
USS *Oklahoma* (BB-37) Memorial, Honolulu, HI 21
USS *Pampanito* (SS-383), San Francisco, CA 14
USS *Razorback* (SS-394), North Little Rock, AR 12

USS *Salem* (CA-139), Fall River, MA 30
USS *Scorpion* (SSN-589) Memorial, North Little Rock, AR 12
USS *Silversides* Submarine (SS-236) Museum, Muskegon, MI 32
USS *Snook* (SS-279) Memorial, North Little Rock, AR 12
USS *Texas* (BB-35), Galveston, TX 47
USS *Utah* (BB-31/AG-16) Memorial, Honolulu, HI 21
USS *Yorktown* (CV-10), Mt. Pleasant, SC 43
Varnum Memorial Armory Museum, East Greenwich, RI 43
Vermilion County War Museum, Danville, Illinois 24
Vermont National Guard Library and Museum, Colchester, VT 49
Veterans Memorial Island Sanctuary, Vero Beach, FL 18
Vietnam Veterans Memorial, Washington, DC 55
Vietnam War Exhibit and Education Center, Melbourne, FL 18
Vietnam War Memorial, Indianapolis, IN 25
Virginia War Museum, Newport News, VA 50
War Eagles Air Museum, Santa Teresa, NM 38
Warhawk Air Museum, Nampa, ID 21
Warren ICBM & Heritage Museum, Francis E. Warren AFB, WY 54
Webb Military Museum, Savannah, GA 19
West Haven Veterans Museum and Learning Center, West Haven, CT 16
Western Sky Aviation Warbird Museum, St. George, UT 48
Wings Over The Rockies Air and Space Museum, Denver, CO 15
Wilson County Veterans Museum, Lebanon, TN 45
Wisconsin National Guard Museum, Camp Douglas, WI 53
Wisconsin Veterans Museum, Madison, WI 53
World War II Glider and Military Museum, Iron Mountain, MI 32
Wright Museum of WWII, Wolfeboro, NH 36
WWII Flight Training Museum, Douglas, GA 20
WWII Home Front Museum, St. Simons Island, GA 20
Wyoming National Guard Museum, Cheyenne, WY 54
Yankee Air Museum, Belleville, MI 32

NOTES

NOTES

NOTES

NOTES

NOTES

NOTES

NOTES